DKfindout!

Science

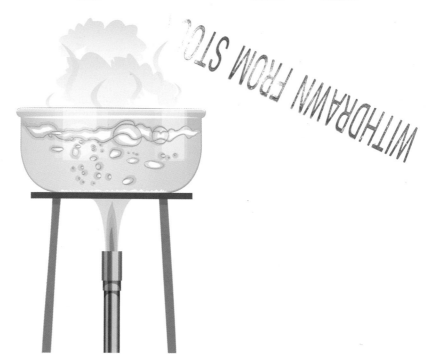

Author and consultant: Dr Emily Grossman

Senior editor Gill Pitts
Project art editor Fiona Macdonald
Editor Katy Lennon
Designer Hoa Luc
Managing editor Laura Gilbert
Managing art editor Diane Peyton Jones
Senior picture researcher Rob Nunn
Pre-production producer Nikoleta Parasaki
Producer Srijana Gurung
Art director Martin Wilson
Publisher Sarah Larter
Publishing director Sophie Mitchell

Educational consultant Jacqueline Harris

First published in Great Britain in 2016 by
Dorling Kindersley Limited
80 Strand, London, WC2R 0RL

A CIP catalogue record for this book
is available from the British Library.
ISBN: 978-0-2412-2519-6

Printed and bound in China

A WORLD OF IDEAS:
SEE ALL THERE IS TO KNOW

www.dk.com

Contents

Simple machines

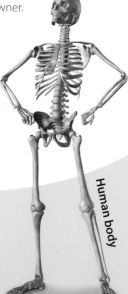

Human body

Reversible change

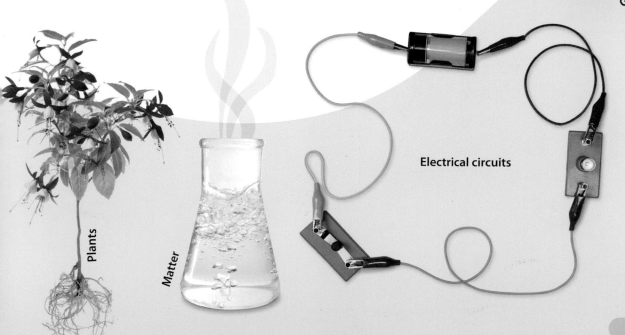

Gravity

Electrical circuits

Plants

Matter

What is science?

Science isn't just a lot of facts in a book. Science is a way of thinking. Science is about asking questions and finding out the answers. Scientists come up with new ideas, and invent new ways to make life easier for everyone. Here are some examples of things that can be explained by science.

Where does electricity come from?

What materials conduct electricity?

Bright sparks
Electricity is one of several different types of energy that have been studied by scientists. Electrical energy flows through wires, and is used to power things like lights and computers.

Where does lightning come from?

Why do plants need sunshine?

Why do humans have bones?

Life on Earth

Cats aren't the only curious animals that live on Earth! Scientists have already discovered nearly 9 million different animals, plants, and other livings things, and they think that there may be many more.

How do you know if something is alive?

Science helps us to understand the world

Material matters

Everything around you is made of matter, and that includes your body! Scientists have explained how matter can take different forms, and why it behaves differently when it is a solid, liquid, or gas.

What is gas?

What is the hardest material on Earth?

What is a magnet?

Forces of attraction

Magnets produce an invisible pulling and pushing force called magnetism. The Ancient Greeks knew this force existed, but it took until 1279 for a scientist to explain how it works!

Why do only some things stick to magnets?

Where does light come from?

How are shadows made?

Casting shadows

Have you ever wondered why you only sometimes have a shadow, and why its length changes? Well, science has the answer, and once you have read this book, you will know the answer too!

around us and how everything works.

What is matter?

Matter is the "stuff" that all things are made of. Everything we see, touch, and breathe is made of matter. Matter has different forms, called states. The three most common states on Earth are solids, liquids, and gases. Matter behaves differently in each of these states.

SOLID

A solid is something that holds its shape, and stops you from moving through it. For example, a rock, a spoon, or a book.

» Solids can be cut or shaped.

» Most solids are hard, like metal.

LIQUID

Liquids do not hold their shape on their own. They will take on the shape of their container, like water in a glass.

» Liquids flow, like this river, and they can be poured, like milk and juice.

» Liquids fall down, as they are pulled towards the Earth by gravity.

Matter matching

Each image below shows an everyday object. Can you work out whether each one shows a solid, liquid, or gas? Read the descriptions for clues.

1 SOLID   **2** LIQUID   **3** GAS

Steam
This steam is rising from the surface of a very hot cup of coffee.

A

B

Backpack
This backpack is made of fabric and strong plastic.

Balloon
This balloon has been filled with air so that it will float. What is the state of the matter inside it?

C

Chair
This chair is made out of wood and has a wicker seat, so it is strong but also comfortable.

D

Sesame oil
Sesame oil is used for cooking and can be poured from a bottle.

E

Switching states

Most things change from one state to another when their temperature changes. When a solid gets hot, it melts and changes into a liquid. When this liquid is heated, it evaporates, changing into a gas or vapour.

Ice is a solid. It is very cold.

When ice is heated above 0°C (32°F), it melts and becomes a liquid called water.

When water is heated to 100°C (212°F), it boils and changes into a vapour called steam.

What is a material?

Some materials are found naturally on Earth, others are made by humans. One material may be hard and brittle, while another may be soft and bendy. These types of features are called "properties". We choose a material for a specific job based on its properties. For example, you wouldn't build a castle out of chocolate!

① Metal

Metals are usually strong, but not brittle, which means they do not break easily. Iron, for example, may be used to make strong chains. Metals are also good at allowing heat and electricity to pass through them.

Types of metal

Different metals have different properties and so are used to make different things.

Gold is a shiny metal from which crowns and jewellery are made.

Iron is heavy. Heated flatirons were once used to press clothes.

Aluminium is a light metal. It is used to make cans and aircraft.

Copper is easy to bend. It is often used for water pipes or wires.

2 Brick

Bricks are rectangular blocks of clay mixed with sand. They are heated to a high temperature to make them very hard. Because we can make as many bricks as we want that are exactly the same size, they are an ideal building material.

3 Glass

Glass is made by melting sand. It is transparent, which means we can see through it, so it is often used to make windows. It can also be made in different colours. Glass is hard, but brittle and so can easily be broken.

4 Wood

Wood comes from the trunks of trees. It is strong and light. Wood is easy to cut into different shapes, so it is used to make bridges, buildings, and furniture. Wood is also easy to set on fire. Burning wood is useful because it produces heat.

5 Cloth

Cloth can be made by knitting or weaving cotton, wool, or silk. It is a soft, light material, and people have made clothes from it for thousands of years. It can also be used to make coloured flags.

6 Rock

Rock is a natural material. For example, mountains are made of rock. Some rocks are hard, but others are soft and crumbly, such as chalk. Humans have often used hard rocks to build tall, strong walls.

Lichtenstein Castle
This castle on top of a cliff near Honau in Germany is made from lots of different materials.

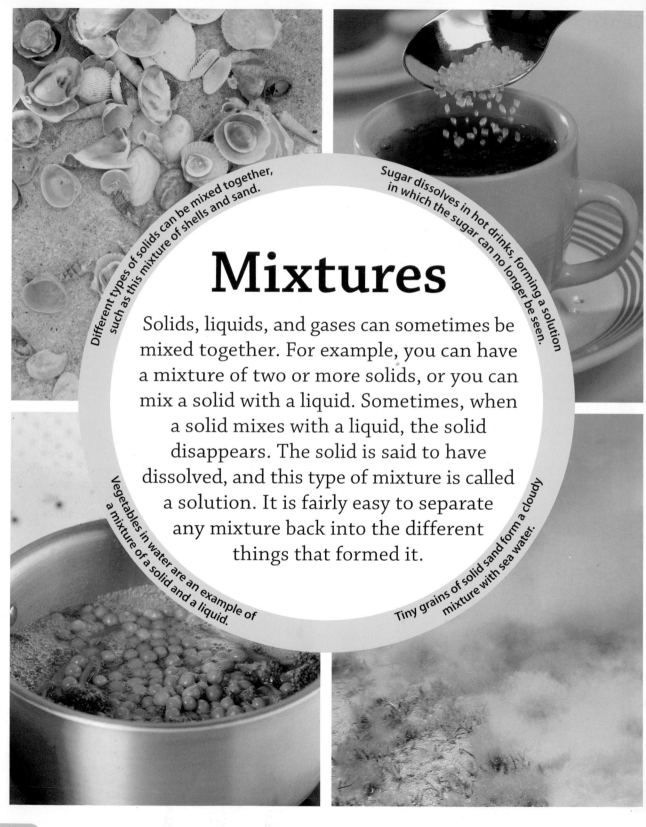

Mixtures

Solids, liquids, and gases can sometimes be mixed together. For example, you can have a mixture of two or more solids, or you can mix a solid with a liquid. Sometimes, when a solid mixes with a liquid, the solid disappears. The solid is said to have dissolved, and this type of mixture is called a solution. It is fairly easy to separate any mixture back into the different things that formed it.

Different types of solids can be mixed together, such as this mixture of shells and sand.

Sugar dissolves in hot drinks, forming a solution in which the sugar can no longer be seen.

Vegetables in water are an example of a mixture of a solid and a liquid.

Tiny grains of solid sand form a cloudy mixture with sea water.

Separating mixtures

Mixtures can be separated back into their different parts. Choosing the correct method depends on what type of things the mixture is made up of.

Sieving

Sieving can be used to separate a mixture of solids, but the solids must be different sizes. Liquids containing large solids can also be separated by sieving.

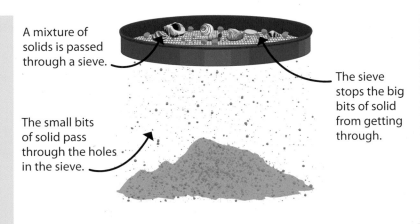

A mixture of solids is passed through a sieve.

The sieve stops the big bits of solid from getting through.

The small bits of solid pass through the holes in the sieve.

Filtering

Mixtures of solids and liquids can be separated by a process called filtration. The mixture is passed through a fine mesh, such as filter paper, which holds back the solid bits, but allows the liquid through.

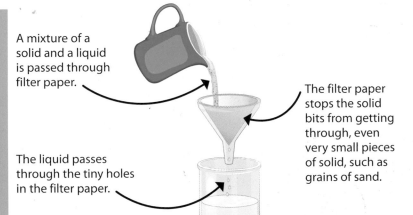

A mixture of a solid and a liquid is passed through filter paper.

The filter paper stops the solid bits from getting through, even very small pieces of solid, such as grains of sand.

The liquid passes through the tiny holes in the filter paper.

Evaporating

When a solid has dissolved in a liquid to form a solution, it cannot be separated by filtration. The solution must be heated until the liquid evaporates, leaving the solid behind.

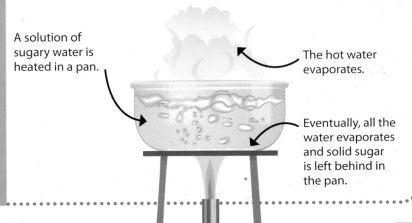

A solution of sugary water is heated in a pan.

The hot water evaporates.

Eventually, all the water evaporates and solid sugar is left behind in the pan.

Changes

Changes are taking place in the world around us all the time. Sometimes, after a change, things can be put back to the way they were before. Other types of change leave things altered forever.

Reversible change

Reversible changes are easy to reverse, or undo. For example, ice can melt to become water, and then freeze, turning back into ice again.

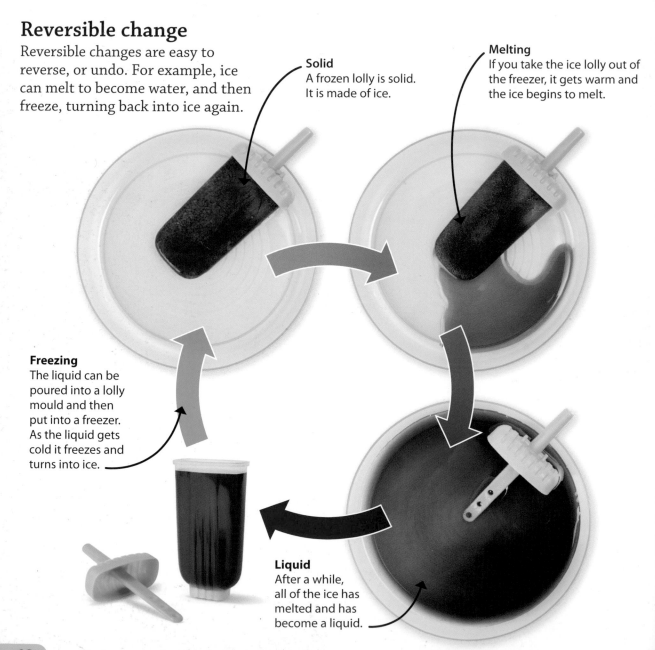

Solid
A frozen lolly is solid. It is made of ice.

Melting
If you take the ice lolly out of the freezer, it gets warm and the ice begins to melt.

Freezing
The liquid can be poured into a lolly mould and then put into a freezer. As the liquid gets cold it freezes and turns into ice.

Liquid
After a while, all of the ice has melted and has become a liquid.

Irreversible change

Irreversible changes cannot be undone. They are permanent. For example, once you have cooked an egg, it cannot be turned back into a raw one!

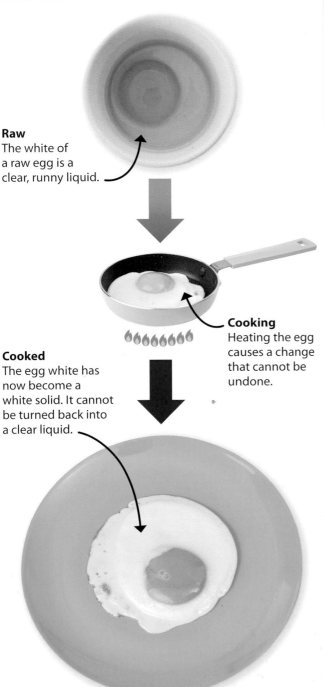

Raw
The white of a raw egg is a clear, runny liquid.

Cooking
Heating the egg causes a change that cannot be undone.

Cooked
The egg white has now become a white solid. It cannot be turned back into a clear liquid.

Everyday changes

Here are some examples of common reversible and irreversible changes that you may see from time to time.

Steamed window
When invisible water vapour in the air hits a cold window, it condenses, turning into tiny water droplets. When the window gets warm, the droplets turn back into water vapour.

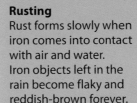

Autumn leaves
Most trees lose their leaves in autumn. Before the leaves fall off the tree, they change from green to red to brown. This change cannot be undone, which means it is permanent.

Rusting
Rust forms slowly when iron comes into contact with air and water. Iron objects left in the rain become flaky and reddish-brown forever.

Rotting food
When food gets old, it can be attacked by tiny living things called mould and bacteria. As the food rots it turns brown, smells nasty, and shrivels up. It cannot be changed back into its fresh form again.

The water cycle

Almost three-quarters of the Earth's surface is water. Water is found in rivers, lakes, and oceans, and it can also be seen as tiny water droplets in clouds, or falling to Earth as rain or snow. Water is constantly moving from one place to another around the planet. This movement is called the water cycle.

The Sun's rays warm the land and the oceans.

Water on the move
The constant cycle of evaporation and condensation of water is almost entirely caused by heat from the Sun.

Evaporation
Heat from the Sun causes water in the ocean to evaporate, turning into invisible water vapour. Water also evaporates from rivers and lakes.

Condensation
As water vapour rises it cools, and condenses into tiny water droplets, forming white clouds.

Atacama Desert, Chile

Mawsynram, India

Clouds are blown across the land by the wind.

As clouds rise and cool, the tiny water droplets come together and fall as rain or snow.

Rain runs over the surface of the land and collects in streams and lakes.

Rivers flow into the ocean.

Snow from the mountains melts and forms streams.

Streams join to form larger rivers, which flow downhill.

Groundwater eventually runs into the ocean.

Water sinks into the ground, forming groundwater beneath the surface of the Earth.

Forces

A force is a push or a pull. Forces are needed to make things move. They can also make things speed up, slow down, or stop moving. Forces can also cause things to change direction or change shape. Most forces only act when things touch each other. Other forces, such as gravity and magnetism, act at a distance.

Pushing forces

A sail boat uses the pushing force of the wind to start moving and to keep moving in the water. If the wind dies down, the boat will slow down and eventually stop, because the air and water are pushing back against it.

Wind
The wind pushes against the sail, moving the boat forwards through the water.

Pulling forces

A small tugboat is able to pull a much larger ship through water because it has a very powerful engine. The engine produces a strong pulling force, which acts through the towline connecting the two boats.

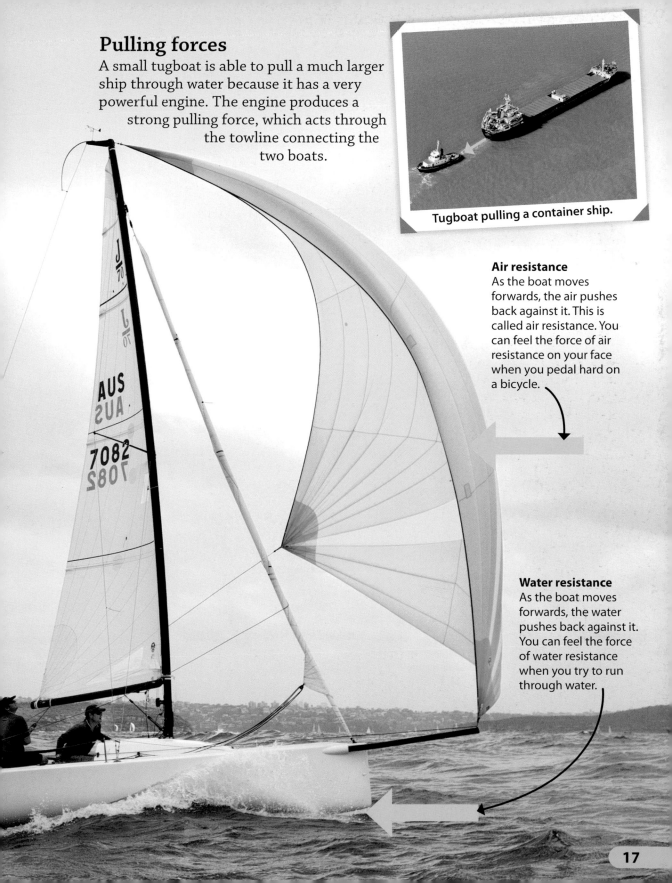

Tugboat pulling a container ship.

Air resistance

As the boat moves forwards, the air pushes back against it. This is called air resistance. You can feel the force of air resistance on your face when you pedal hard on a bicycle.

Water resistance

As the boat moves forwards, the water pushes back against it. You can feel the force of water resistance when you try to run through water.

AUS
7082

Laws of Motion

Nothing can move without a force to start it on its way. Forces are also needed to make things move faster, to change their direction, and to make them stop. In 1687, the English scientist Isaac Newton came up with three simple ideas that show how forces affect the way things move. These ideas are called the Laws of Motion.

This ball will stay where it is until it is kicked, or blown by a strong gust of wind.

The kick provides a pushing force on the ball.

Newton's First Law

If an object is not moving, it will stay completely still unless an outside force acts on it. If an object is already moving, it will keep on moving at the same speed, and in a straight line, unless an outside force, such as gravity, causes it to change its motion.

Moving forever

Every object that is moving through air will eventually come to a stop, because the air pushes back against the moving object, slowing it down. This force is called air resistance. However, in space there is no air, which means that once something starts to move, it will keep moving forever! This is an example of Newton's First Law.

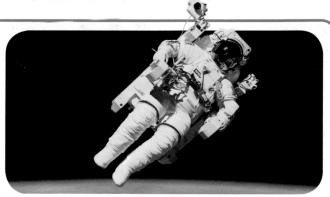

An astronaut floating through space.

The ball moves off in the same direction as the force.

When this ball hits the wall, it pushes against the wall in the direction it was moving in.

The wall pushes back on the ball with the same amount of force, and the ball bounces off in the opposite direction.

Newton's Second Law

If an outside force pushes or pulls on an object that is not moving, it will start to move in the same direction as the force. If an object is already moving, an outside force will cause it to move faster, slow down, or change direction.

Newton's Third Law

This law says that "Every action has an equal and opposite reaction". This means that when a force acts in one direction, it creates another equal force in the opposite direction. An example of this is a ball bouncing off a wall.

Friction

Friction is a force that slows things down. It occurs when two surfaces rub against each other. Different types of surface produce different amounts of friction. Smooth surfaces, such as ice, do not create much friction, so things slide over them easily. Rough surfaces, such as roads, produce a lot of friction. This helps car wheels to grip the road, and allows cars to stop easily.

Sliding race
Four metal bobsleighs are held at the top of a slope. Each slope has a different surface. The bobsleighs are let go of at the same time. Which one will move the fastest?

Super friction

If the pages of two books are interleaved, it is almost impossible to pull them apart. This is because there are hundreds of pages, and therefore hundreds of surfaces, that are trying to slide against each other at the same time. This produces an enormous amount of friction, which stops the pages from being pulled apart.

Linked books
The pages of these two books have been alternately stacked.

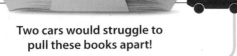

Two cars would struggle to pull these books apart!

Slippery surface

A smooth surface, such as this slope covered in oil, creates hardly any friction. Objects slide very easily and quickly along them. Ice skaters and skiers make use of slippery surfaces.

Wooden floorboards

Floorboards are fairly smooth, producing only a little friction, which allows objects to slide over them quite easily. This is why you can sometimes slide on floorboards in your socks.

Scratchy sandpaper

Sandpaper is rough and so creates quite a lot of friction. This means that an object sliding over sandpaper will slow down and stop more quickly than an object sliding over a smoother surface.

Green grass

A very rough surface like grass produces lots of friction, causing an object sliding over it to slow down a lot. The longer the grass, the more quickly the object will slow down.

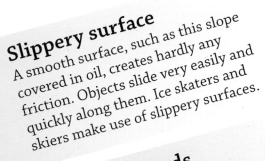

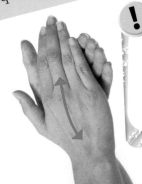

REALLY?

When **you rub your hands together**, it is **friction** that causes them to **get warm**.

Gravity

Gravity is an invisible force that pulls things towards the centre of the Earth. It pulls everything towards the Earth, whether the object is in the air, in water, or already on the ground. Even the Moon is kept moving around the Earth by the pull of Earth's gravity.

Leaping squirrel

Gravity keeps things held to the Earth. For something to be able to lift off the ground, it has to produce a force large enough to overcome the force of gravity.

Upwards force

Gravity

Gravity

1 Jumping up
When the squirrel jumps, its legs push against the branch, producing an upwards force. This force is greater than the force of gravity, so the squirrel lifts off the branch.

2 Gaining height
There is no longer a force from the legs. The squirrel keeps moving upwards, but the force of gravity slows it down.

The Moon

The Moon has its own gravity, but because it is smaller than the Earth, its force of gravity is not so strong. The squirrel would feel lighter on the Moon, and it would be able to jump almost seven times higher.

Blast off!
A huge explosion is needed to produce enough force to overcome the force of Earth's gravity and launch a rocket into space.

The squirrel takes a rocket to the Moon.

Gravity

3 Falling down
The squirrel slows down so much that for a tiny moment it stays totally still in the air. Then gravity causes the squirrel to begin to fall towards the ground.

4 Landed
Once the squirrel has landed on another branch, the force of gravity holds it there until it decides to jump again.

Gravity

Simple machines

Machines are used to make a task easier. They reduce the amount of effort that we need to put in (the force) to lift or move a heavy object (the load). Machines can also change the direction of a force, so we can move things in a more helpful way.

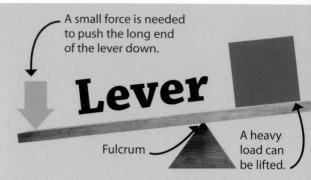

A small force is needed to push the long end of the lever down.

Lever

Fulcrum

A heavy load can be lifted.

A lever is a bar that swivels on a fixed point, called a fulcrum, and makes it easier to lift a heavy load. If you push down at one end of the lever, the load at the other end is raised.

Gears

Gears are wheels with teeth that fit together. When one gear is turned, the other one turns in the opposite direction. If the gears are different sizes, they turn at different speeds.

The larger gear turns more slowly.

Tooth

The smaller gear turns more quickly.

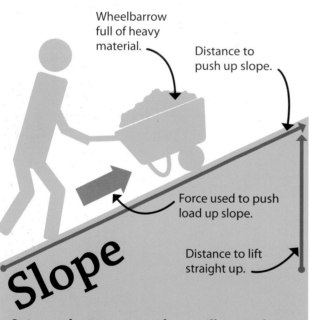

Wheelbarrow full of heavy material.

Distance to push up slope.

Force used to push load up slope.

Distance to lift straight up.

Slope

It is much easier to push or pull something heavy up a slope than it would be to lift it straight up because less force is needed. However, you have to push or pull the object a greater distance to get it to the same place.

Wedge

A wedge helps us to push things apart. The blade of an axe is a type of wedge. When the axe is swung and the blade hits a piece of wood, the wedge forces the wood to split apart.

Axe blade splits the wood between its fibres.

Pulley

A pulley makes it easier to lift a heavy object (load) straight up. A rope is looped over a wheel and one end of the rope is attached to the load. It is easier to lift the load by pulling down on the rope than by picking it up.

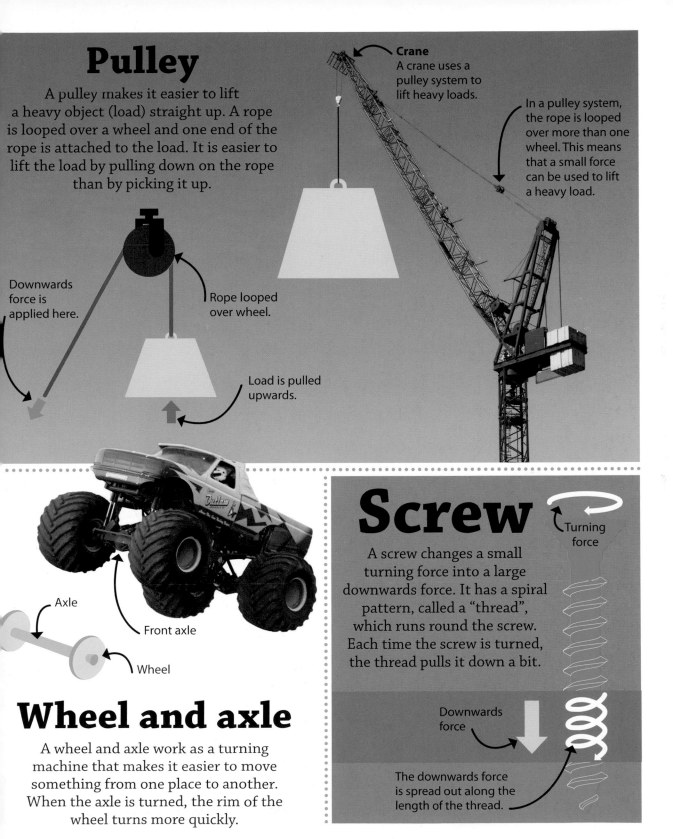

Crane
A crane uses a pulley system to lift heavy loads.

In a pulley system, the rope is looped over more than one wheel. This means that a small force can be used to lift a heavy load.

Downwards force is applied here.

Rope looped over wheel.

Load is pulled upwards.

Screw

A screw changes a small turning force into a large downwards force. It has a spiral pattern, called a "thread", which runs round the screw. Each time the screw is turned, the thread pulls it down a bit.

Turning force

Downwards force

The downwards force is spread out along the length of the thread.

Axle

Front axle

Wheel

Wheel and axle

A wheel and axle work as a turning machine that makes it easier to move something from one place to another. When the axle is turned, the rim of the wheel turns more quickly.

Energy

Energy is needed to make things happen. Every movement or change, no matter how small, requires energy. Energy has many different forms. For example, some types of energy are needed to move cars or light up homes. Your body needs energy to move, grow, and keep warm.

Electrical

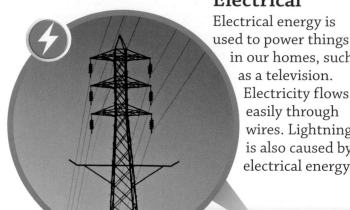

Electrical energy is used to power things in our homes, such as a television. Electricity flows easily through wires. Lightning is also caused by electrical energy.

Movement

All moving things have energy, from dogs to waterfalls. The faster something moves, the more energy it has.

! WOW!

Without energy our world would be **freezing cold**, dark, silent, and **lifeless.**

Chemical

There is energy stored inside the chemicals in food. When animals eat, their body breaks down the food, releasing the energy.

Light

Glowing objects give out light energy, which we see with our eyes. Nearly all the energy on Earth comes originally from the Sun.

Stored energy

Sometimes energy is trapped inside things. This is called stored, or potential, energy. When this stored energy is released, it can make things happen. A stretched catapult stores elastic energy. When you release it, it gives energy to the ball, making the ball fly out.

The more the catapult elastic is stretched, the more stored energy it has.

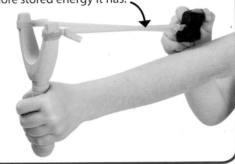

Sound

Sound is a form of energy that is produced when objects vibrate, or shake. For example, when a bell is hit, it makes a sound. We receive sound energy through our ears. This is called hearing.

Heat

Hot things contain lots of heat energy. They pass on this energy to cooler things around them, so hot things like a fire can be used to keep us warm. We get most of our heat energy from the Sun, and from burning things, such as wood.

Heat

Heat is a type of energy that we can feel. The hotter something is, the more heat energy it gives out. Heat energy always tries to spread from a hotter to a colder thing. So, when you touch a hot object, heat energy flows out of the object into you, warming you up.

Convection

When gases and liquids get hot, heat energy spreads out through them by convection. Hot air or water rises, and cold air or water sinks to take its place.

Cool air sinks.

Smoke produced by the burning wood.

Conduction

Heat energy spreads out through solid objects by conduction. Metal objects are very good conductors of heat, so heat energy passes through them quickly and easily.

Hot air rises.

Heat from the fire spreads through the metal pan.

Radiation

Heat escapes from the surface of hot objects by invisible rays that travel through air and space. This is how we feel the heat from the Sun. Hotter objects give out more radiation than cooler ones.

Radiation rays from the hot fire.

Temperature

Temperature is the measure of how hot or cold things are. It is recorded in degrees Celsius (°C) or degrees Fahrenheit (°F).

15 million°C
(27 million°F)
The temperature at the centre of the Sun.

1,760°C
(3,200°F)
The temperature at which sand melts and turns into glass.

100°C
(212°F)
The boiling point of water.

232°C
(450°F)
The temperature at which dry wood catches fire.

56.7°C
(134°F)
The hottest temperature recorded on Earth at Death Valley, California, USA, on 10 July 1913.

37°C
(98.6°F)
The average temperature of the human body.

2°C
(35.6°F)
The temperature of the air when rain turns to snow.

0°C
(32°F)
The freezing point of water.

-89.2°C
(-128.5°F)
The coldest temperature recorded on Earth at Vostock Station, Antarctica, in 1983.

-272°C
(-458°F)
The temperature of the Boomerang Nebula, a cloud of dust and gas that is the coldest place in the Universe.

Sound

Sound is produced when things vibrate, or shake. A vibrating object makes the air next to it vibrate as well. These vibrations travel through the air as a sound wave, which we hear with our ears. Sound waves can also travel through liquids and solids, such as water, rock, or wood.

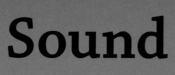

Making sound

When the climber hits the metal spike with his hammer, it vibrates. This vibration causes the surrounding air to vibrate too, producing a sound wave.

Sound wave

The sound wave spreads out through the air. It travels away from the rock and hammer until it reaches a solid surface.

Loudness

Vibrations with lots of energy produce big waves and make loud sounds. So, banging a drum harder will produce a louder sound. Sound waves lose energy as they travel, which means that the further you are from a sound, the quieter it will be.

Reflection

When the sound wave hits a solid surface, such as the side of a cliff, some of the energy bounces back towards the climber. The rest of the sound energy travels through the rock.

Echo

The climber will hear the reflected sound wave as an echo, a few moments after he first produced the sound.

Rustling leaves
When leaves brush against one other, they produce tiny vibrations in the air. These vibrations only have a small amount of energy, so the rustling sound is very quiet.

Small shallow wave

Roaring lion
A lion's roar causes the surrounding air to vibrate with lots of energy. It produces a loud sound, which can be heard up to 8 km (5 miles) away.

Bigger wave

Rocket launch
When a rocket launches, the exploding gases produce huge amounts of sound energy. The noise is louder than 10 million rock bands performing at once!

Large deep wave

Light

Light is a type of energy that allows us to see things. Light travels as rays, in straight lines. Light rays are given off by things that glow, such as the Sun, a candle, or a light bulb. These things are called light sources. Other objects around us look bright because they bounce light from light sources back to our eyes. When there is no light source everything looks dark, for example at night.

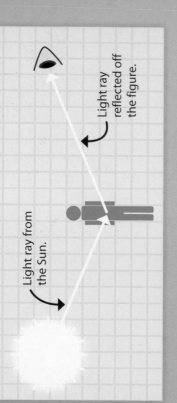

Seeing things

Most things do not give out their own light. We can see these objects when we look at them because they reflect the light that falls on them. Light from a light source, such as the Sun, bounces off the object and into our eyes.

Light ray from the Sun.

Light ray reflected off the figure.

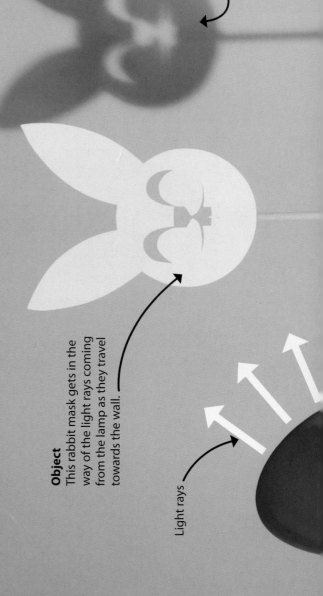

Object
This rabbit mask gets in the way of the light rays coming from the lamp as they travel towards the wall.

Light rays

Shadow
A shadow forms where the light from the lamp has been blocked by the rabbit mask. The light rays have been stopped from reaching the wall.

Light source
This lamp is a light source. It gives out light rays that travel towards the wall in straight lines.

Creating shadows

If something solid gets in the way of light coming from a source, the light rays are blocked. This creates a shadow behind the object, which is the same shape as the object.

Shadow length

The length of the shadow of an object can change, depending on where the light source is. The more directly above an object a light source is, the shorter the object's shadow will be.

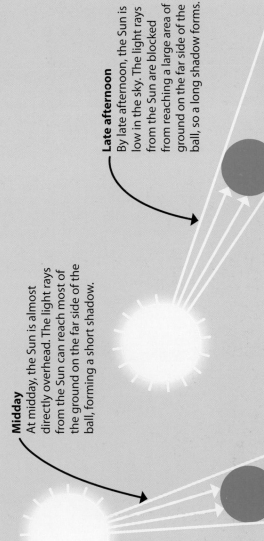

Midday
At midday, the Sun is almost directly overhead. The light rays from the Sun can reach most of the ground on the far side of the ball, forming a short shadow.

Late afternoon
By late afternoon, the Sun is low in the sky. The light rays from the Sun are blocked from reaching a large area of ground on the far side of the ball, so a long shadow forms.

Rainbows

We usually see light as white, but it is actually a mixture of different colours. We know this because we can see the colours in a rainbow. When the Sun is shining low in the sky behind you and rain is falling in front of you, you may see a rainbow.

Seeing a rainbow
As sunlight passes through a shower of raindrops, it bends and splits into different colours.

Double rainbow
If the sunlight is reflected twice inside each raindrop, you can see a second rainbow outside the main one, with its colours in the opposite order.

Inside a raindrop

1 Rays of sunlight shine towards the falling raindrops.

2 When a ray enters a raindrop, it bends and the white light spreads into different colours.

3 The light reflects off the back of the raindrop.

4 When the light leaves the raindrop it bends again, so the colours spread out even more and shine down towards your eyes.

Red

Orange

Yellow

Green

Blue

Indigo

Violet

The colours of a rainbow

The colours always spread out in the same order. We call the pattern they make a spectrum. The seven main bands of colour blend into each other, creating millions of different colours.

Colour arc
Rainbows are curved because the colours leave the raindrops at different angles.

Bright colours
Large raindrops form bright, clear rainbows, while small raindrops make faint, fading ones.

Is it the end?
If the ground wasn't in the way, you would see the rainbow as a circle.

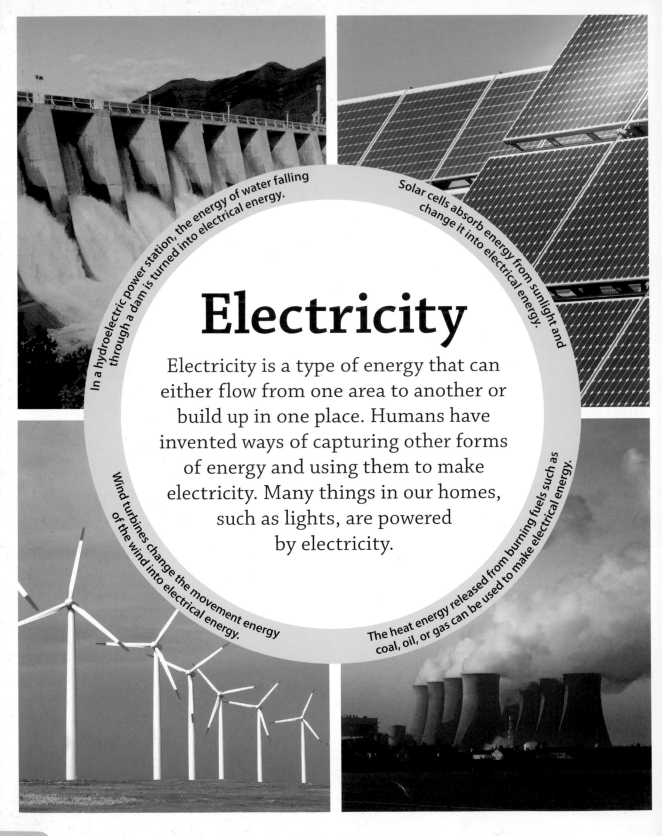

Electricity

Electricity is a type of energy that can either flow from one area to another or build up in one place. Humans have invented ways of capturing other forms of energy and using them to make electricity. Many things in our homes, such as lights, are powered by electricity.

In a hydroelectric power station, the energy of water falling through a dam is turned into electrical energy.

Solar cells absorb energy from sunlight and change it into electrical energy.

Wind turbines change the movement energy of the wind into electrical energy.

The heat energy released from burning fuels such as coal, oil, or gas can be used to make electrical energy.

A series of cables and pylons are used to transport electricity from place to place.

Into our homes

After electricity has been made by a power station, it travels to towns and homes along a series of underground or overground cables. The electrical cables send electricity into the plug sockets found in most buildings. This is called mains electricity.

! WOW!

Electricity travels through wires **100 times slower** than the **speed of light**.

Static electricity

Sometimes electrical energy builds up in one place. This is called static electricity. The build-up can happen when things are rubbed together, like a balloon and a jumper, which can cause them to stick together. When a build-up of electrical energy is released, a spark of electricity can sometimes be seen or felt.

Lightning
A lightning bolt is a giant spark of electricity. When rain clouds rub against the air around them, static electricity builds up. If the build-up gets big enough, the electrical energy leaps to Earth as a bolt of lightning.

Battery

A battery stores electrical energy and releases it when connected to a circuit. The battery will eventually run down and will have to be replaced.

Switch

A switch allows the bulb to be turned on or off without removing the battery. When the switch is closed, the circuit is complete and the bulb lights up.

Electrical circuits

An electrical circuit is a loop around which electrical energy can move. Electricity travels from a power source, such as a battery, through wires, and back to the power source. In a simple circuit, other useful items that make use of the electricity, such as a bulb or a buzzer, are added. Electricity can only flow when there is a complete circuit.

Open switch

When the switch is open, there is a gap in the circuit and the bulb cannot light up.

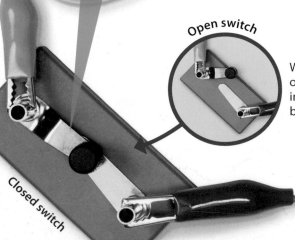

Closed switch

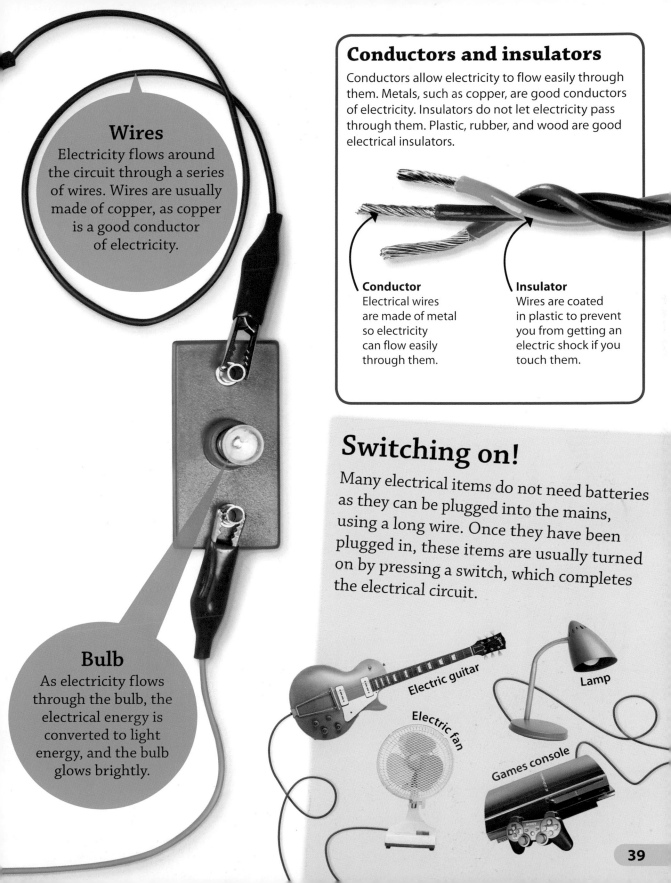

Wires

Electricity flows around the circuit through a series of wires. Wires are usually made of copper, as copper is a good conductor of electricity.

Conductors and insulators

Conductors allow electricity to flow easily through them. Metals, such as copper, are good conductors of electricity. Insulators do not let electricity pass through them. Plastic, rubber, and wood are good electrical insulators.

Conductor
Electrical wires are made of metal so electricity can flow easily through them.

Insulator
Wires are coated in plastic to prevent you from getting an electric shock if you touch them.

Bulb

As electricity flows through the bulb, the electrical energy is converted to light energy, and the bulb glows brightly.

Switching on!

Many electrical items do not need batteries as they can be plugged into the mains, using a long wire. Once they have been plugged in, these items are usually turned on by pressing a switch, which completes the electrical circuit.

Electric guitar

Lamp

Electric fan

Games console

Magnetic materials

Iron is a magnetic material, so any metal with iron in it, such as steel, will be attracted to a magnet. Nickel and cobalt are also magnetic metals.

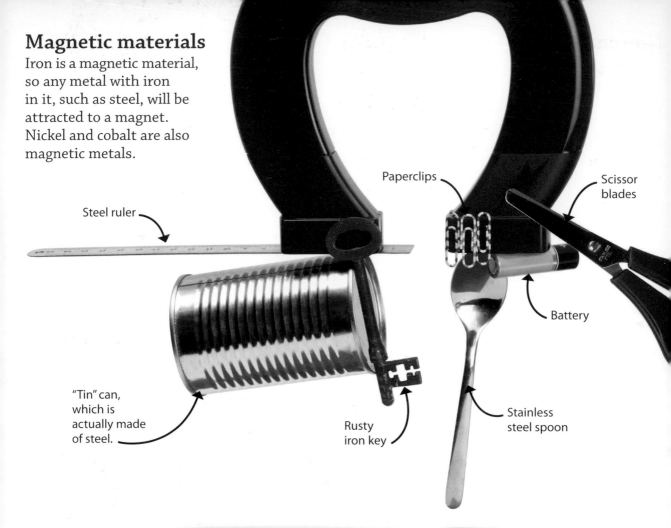

Steel ruler

Paperclips

Scissor blades

Battery

"Tin" can, which is actually made of steel.

Rusty iron key

Stainless steel spoon

Magnets

Magnets are objects that produce an invisible force called magnetism. When two magnets are close to each other they produce pushing or pulling forces on one another. Magnets can also pull on, or attract, other objects that are made of magnetic materials, such as iron. Magnets are often U-shaped, like a horseshoe.

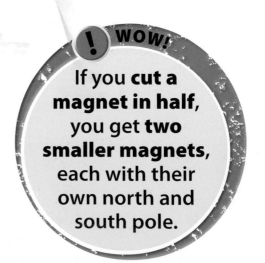

! WOW!

If you **cut a magnet in half**, you get **two smaller magnets**, each with their own north and south pole.

Non-magnetic materials

Magnets have no effect on non-magnetic materials. Non-magnetic materials include all non-metals, such as glass, plastic, or wood. Most metals are also non-magnetic, including gold, aluminium, and copper.

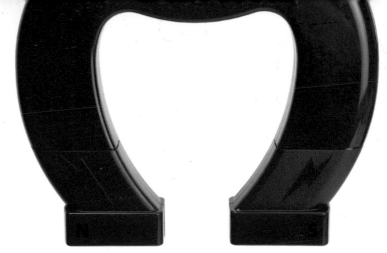

Plastic scissor handles

Rubber dinosaur

Silver fork

Wood blocks

Glass bottle

Aluminium can

Woolly pompoms

Lemon

Magnetic poles

Magnets have two ends, or poles, called a north pole and a south pole. Unlike poles pull on each other, or attract, so a north pole pulls a south pole. Like poles push on each other, or repel, so a north pole pushes another north pole away.

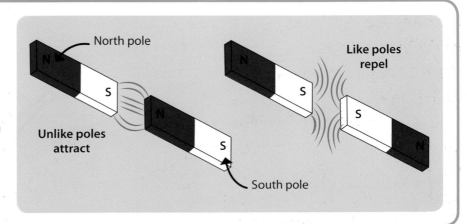

North pole

Like poles repel

Unlike poles attract

South pole

Meet the expert

Dr Suze Kundu is a materials scientist at Imperial College London in the UK. She works on materials that capture energy from the Sun. Dr Kundu also gives talks at schools and science festivals, and is a science writer and television presenter.

Q: We know it is something to do with science, but what do you actually do?

A: I work in a laboratory, finding out what new materials are like and how they act under different conditions. I then find ways in which a new material can be used, and make changes to it, so it works in the best way possible for a specific task.

Q: What made you decide to become a materials scientist?

A: I didn't really decide to become a materials scientist! I followed my heart and studied a subject at university that I really loved, which was chemistry. That was when I discovered nanotechnology, which is working with individual molecules and atoms, and creating new materials. It was the idea of inventing new materials and finding ways to use them that excited me.

Q: What is a usual work day for you?

A: When I am in the laboratory, I work on a material that can capture energy from sunlight and use it to split water. Water is made of hydrogen and oxygen, and so by splitting it we can make

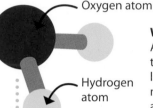

Oxygen atom

Hydrogen atom

Water molecule
A molecule is a group of tiny things called atoms that are linked together. One water molecule has one oxygen atom and two hydrogen atoms.

hydrogen gas, which can be collected and used as a cheap and clean fuel. When hydrogen is burned, lots of energy is released, and water is produced. This means that, for example, hydrogen can be used to power a car, and it will produce water vapour instead of pollution-causing exhaust gases.

Q: What sort of equipment do you use?

A: I use the usual things that you see in laboratories, like glass beakers and flasks. I make solutions that are put on materials in very thin layers, called films. This can be done by spraying or dipping, so I have

Hydrogen cell car
The Toyota Mirai is one of the new types of car that are powered by hydrogen.

equipment that can do that. The films must then be baked, but an oven wouldn't get hot enough, so I use furnaces that can get as hot as 1,000°C (1,830°F)! After that, I use special equipment that tells me what the films look like, and how they act. Finally, I test the films to see how well they split water into hydrogen when white light is shone on them.

Q: What are your days like when you are not in the laboratory?

A: My days outside the laboratory are very varied. I spend a lot of time teaching students. I also write science articles and talks, prepare for shows at schools, do news and radio interviews, and film television programmes. I love the variety in my job, and I wouldn't have it any other way!

Q: What is the most difficult thing about your job?

A: One day an experiment could work really well, and I think I have stumbled across something pretty amazing. Yet, if I try the experiment the next day and don't get the same result, I wonder why. It could be due to the tiniest change. That can be very frustrating, but if I can get the same result again and again, then I know I am on to a winner!

The equipment that Dr Kundu uses includes glass flasks.

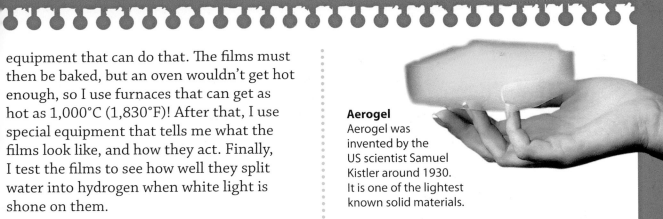

Aerogel
Aerogel was invented by the US scientist Samuel Kistler around 1930. It is one of the lightest known solid materials.

Q: Do you have a favourite material?

A: My favourite material ever invented is aerogel. It is like a solid dry jelly, and can be made of materials similar to glass. It is very light, and has lots of pockets of air inside it, which means heat passes through it very slowly. NASA has used it on space missions to study comets, which are balls of ice and dust that circle the Sun. The aerogel collects bits of dust thrown off the back of the comet when it gets hot.

Q: What do you love most about being a materials scientist?

A: Materials science is the science of all the stuff around us, from the clothes we wear to the food we eat, and the mobile phones we take selfies with. By studying it, I feel that I understand how much science is in all of the things in our lives.

> **❝ I love the fact that my work builds on the history of science discoveries, and that I am helping to make new ones! ❞**

Living things

All living things have certain characteristics in common. To count as being alive, a living thing has to be able to carry out seven different processes, which are listed below.

Life checklist

Movement
All living things can move on their own. Even plants have leaves that turn to face the Sun.

Reproduction
Living things can produce new versions of themselves.

Sensitivity
Living things can detect and respond to changes in the world around them. For example, the ability to react to changes in light, or to hear sounds.

Growth
Living things get bigger as they get older until they reach their full size.

Respiration
All living things turn food into energy, using oxygen from the air.

Excretion
All living things must get rid of any waste that they produce.

Nutrition
All living things need food. Unlike animals, plants make their own food.

Animal

Animals can move from place to place. They cannot make their own food, so they rely on eating other living things to survive.

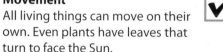

Plant

Plants are fixed in the ground, but their roots, leaves, and flowers can move. They make their own food by using the Sun's rays.

Fungus

Funguses may not look alive, but they are. Most funguses feed on the remains of dead plants or animals. Mushrooms, toadstools, and moulds are all funguses.

Is it alive?

A car can move, and it uses fuel as its food. It turns this fuel into energy, and it gives out its waste as exhaust fumes. However, a car cannot grow or produce baby cars, so it is not alive!

Tree trunk

When a tree dies it usually falls to the ground. Funguses can grow on the dead tree trunk and soak up its nutrients.

! WOW!

There are about **8.7 million** different types of living things on Earth.

Plants

Plants make their own food using sunlight. Most plants are held in the ground by their roots, and have green leaves. There are thousands of types of plants, ranging from huge oak trees to small ones like this fuchsia.

Bees are attracted to sweet-smelling, colourful flowers. While inside a flower, they pick up pollen then pass it to other flowers they visit.

Berries

Fruits, such as berries, are the parts of a flowering plant that contain seeds. Once in the ground, the seeds will grow into new plants.

Flowers

Most plants have flowers for reproduction. They have male pollen and female eggs, which join together to make seeds.

Leaves

Plants use their green leaves to capture energy from the Sun's rays. The leaves use this energy, together with carbon dioxide from the air, and water, to make food for the whole plant.

Stem

The stem supports the leaves and flowers, holding them up towards the light. The stem also carries water and nutrients in the form of mineral salts from the roots to the rest of the plant.

Roots

Plants usually grow in the earth or soil. The roots dig deep into the ground, keeping the plant in place. They also soak up mineral salts and water from the soil.

Why we need plants

Nearly every animal relies on plants in some way or other. Some animals, called herbivores, eat only plants. Humans grow lots of types of plants for food, and also some plants just for their flowers. Some types of tree are grown for their wood.

WOW!

The world's biggest plant is the General Sherman. This giant sequoia tree is 84 m (275 ft) tall.

Spreading seeds

Seeds contain everything needed to form a new plant. To do that, they need to be scattered away from the parent plant to a new place on the soil where they can grow. Different types of plant have different ways of spreading their seeds. Some need animals to move them, while others use wind or even fire to scatter them.

Blown in the wind

Some plants have very light seeds and so can use the wind to spread them. This dandelion has seeds with tiny parachutes that allow them to be carried away when the wind blows.

Yum!

Tasty treats

Many plants have seeds that are hidden inside tasty fruits, such as berries. When an animal eats the fruit, the seeds pass through its body unharmed and are released in its droppings.

Sticky seeds

Animals can carry seeds away from plants without even realizing it. Sticky seeds will fix on different parts of their body, while some seeds, called burs, have little hooks that attach to the animal's fur.

Earth, wind, and fire

Some plants have more dramatic or unusual ways of spreading their seeds. Fire may kill the parent plants, but it leaves behind fertile ash for their seeds to grow in.

Heat treatment
Jack pine cones are glued shut with resin. When a fire sweeps through the trees, the resin melts and the seeds are released.

Desert rover
Tumbleweed is the dried-up top part of some plants. It rolls around the desert whenever the wind blows, scattering its seed as it goes.

Exploding pods

The Himalayan balsam plant keeps its seeds in pods. These pods explode when they are ripe, shooting the seeds out of them. The explosion can scatter the seeds up to 7 m (22 ft) away, often startling unsuspecting passers-by.

Pop!

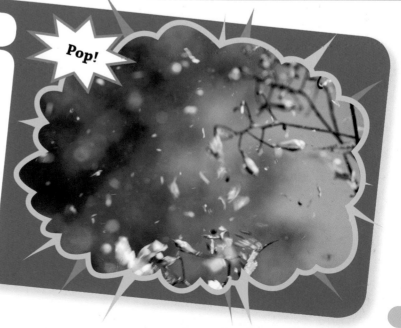

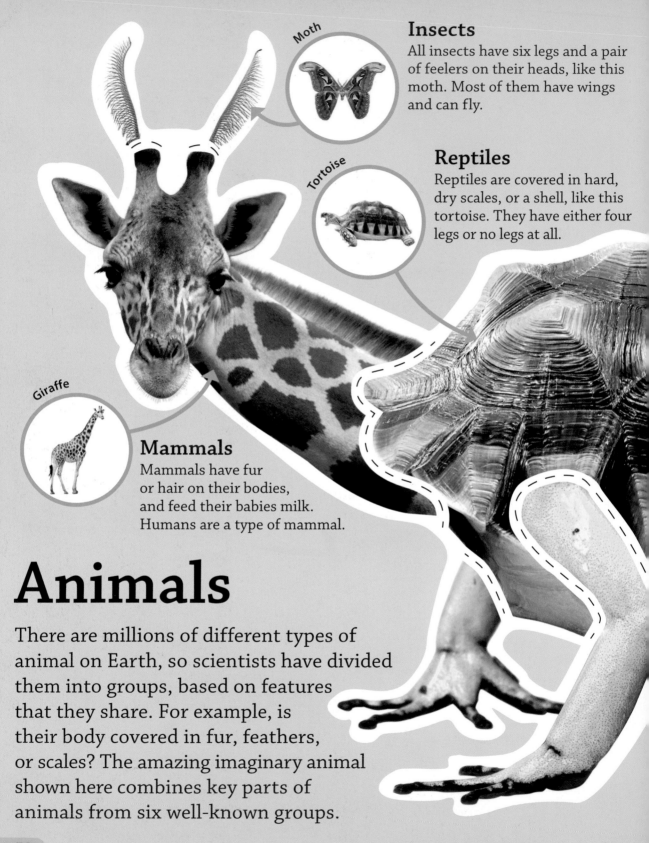

Insects

All insects have six legs and a pair of feelers on their heads, like this moth. Most of them have wings and can fly.

Moth

Reptiles

Reptiles are covered in hard, dry scales, or a shell, like this tortoise. They have either four legs or no legs at all.

Tortoise

Giraffe

Mammals

Mammals have fur or hair on their bodies, and feed their babies milk. Humans are a type of mammal.

Animals

There are millions of different types of animal on Earth, so scientists have divided them into groups, based on features that they share. For example, is their body covered in fur, feathers, or scales? The amazing imaginary animal shown here combines key parts of animals from six well-known groups.

Birds

Birds have wings, and they are the only animals that have feathers, which keep them warm and help them to fly.

Parakeet

Fish

All fish live in water and use their tails to help them swim. Their bodies are covered in scales, and they have gills for breathing underwater.

Mandarinfish

Amphibians

Amphibians live both on land and in water. Most of them have four legs, which they use for walking and swimming.

Frog

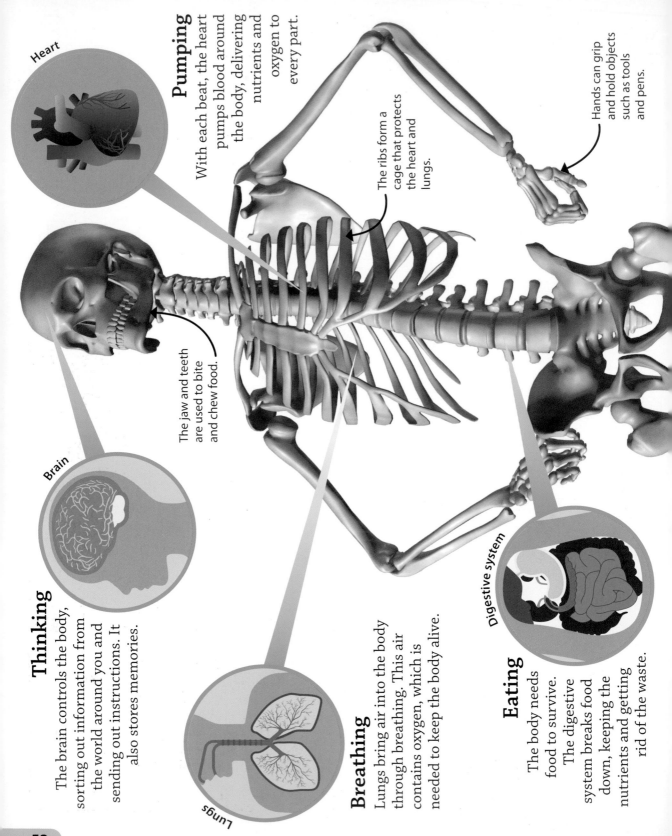

Pumping

With each beat, the heart pumps blood around the body, delivering nutrients and oxygen to every part.

Heart

The ribs form a cage that protects the heart and lungs.

Hands can grip and hold objects such as tools and pens.

The jaw and teeth are used to bite and chew food.

Thinking

The brain controls the body, sorting out information from the world around you and sending out instructions. It also stores memories.

Brain

Breathing

Lungs bring air into the body through breathing. This air contains oxygen, which is needed to keep the body alive.

Lungs

Eating

The body needs food to survive. The digestive system breaks food down, keeping the nutrients and getting rid of the waste.

Digestive system

Being human

Like all vertebrate animals, humans have a bony skeleton underneath their skin and muscles. It is a framework to hold up the body, help it move, and protect what is inside. The amazing brain makes humans the cleverest of all the animals. That includes you!

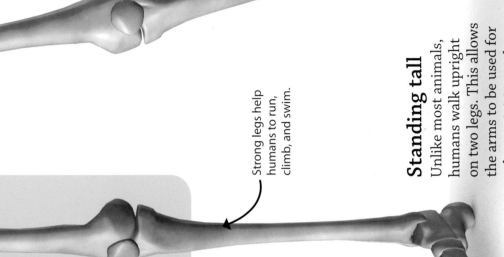

Strong legs help humans to run, climb, and swim.

Strong muscles

Muscles are like elastic straps that can stretch or squeeze. Many muscles move the body by pulling on the bones.

Standing tall

Unlike most animals, humans walk upright on two legs. This allows the arms to be used for other activities such as making things.

4 Body facts

1 Human skeletons contain more than 200 bones. The smallest bone is inside the ear and is only 3 mm ($^3/_{25}$ in) long.

2 The thigh bone is the strongest bone in the body. It is about four times stronger than concrete.

3 A human heart beats more than 100,000 times in a day. That is over 35 million times in a year.

4 In one day, blood travels about 19,000 km (12,000 miles) around the body.

Charles Darwin, 1809–1882, UK

Moth

Beetle

Evolution

Charles Darwin worked out how animals, such as moths and beetles, can change over many generations to become new species. This process is called evolution.

Great scientists

Life is full of problems, and scientists are always trying to find new ways to solve them. There are hundreds of great scientists who have changed our lives for the better, through their discoveries and their inventions. Here are seven of them.

Computers

Ada Lovelace wrote the first published computer program. She also predicted that a computer would be able to make music and images, not just do sums.

Ada Lovelace, 1815–1852, UK

Light bulb

Light bulb

Thomas Edison is best known for inventing the first light bulb that could be made in large numbers. He also invented a sound-recorder and a moving-image projector, which helped to start the age of movies.

Thomas Edison, 1847–1931, USA

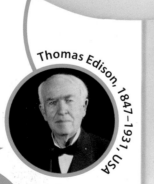

"To invent, you need a good imagination and a pile of junk."

Modern computer

Gravity

Isaac Newton is said to have discovered how gravity works when he saw an apple fall from a tree. He realized that there must be a similar force that keeps the Earth moving around the Sun.

Isaac Newton, 1642–1727, UK

Radioactivity

Marie Curie discovered two substances, called radium and polonium, which give off invisible rays that can pass through materials. She called these rays radioactivity.

Marie Curie, 1867–1934, Poland

Marie Curie's work also helped with the development of X-rays for use in operations in hospitals.

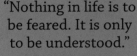
"Nothing in life is to be feared. It is only to be understood."

Relativity

Albert Einstein's theories of relativity, along with his ground-breaking $E=mc^2$ equation, helped scientists to understand the Universe, and how energy, mass, space, and time are all related.

Albert Einstein, 1879–1955, Germany

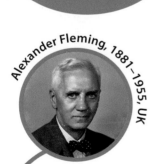
Alexander Fleming, 1881–1955, UK

Antibiotics

Alexander Fleming discovered penicillin, which led to the creation of a group of medicines called antibiotics. They kill the bacteria that cause many infections in humans and other animals, and so have saved millions of lives.

Antibiotics

01010001010
01101010100
00001010100
00100100100
01101010101
10111100100
00010101010
01011100000
00101000100
00010010010
10001010000
11010101000
01010011110
00010101010

Einstein's famous equation

Science in action

Science is all around us. From the moment we wake up to the moment we go to bed, almost everything that we do or that happens to us can be explained by science. Here are a few examples of the ways that science helps us to understand what goes on in our daily lives.

Wakey wakey!

Science explains how a cockerel could wake us up in the morning. Sound waves travel through the air from the cockerel's mouth to our ears.

Cock-a-doodle-do!

Light the way

Science explains how flicking a switch turns on our lights, by allowing electricity to flow around an electrical circuit.

Keeping warm

Science explains how our clothes keep us warm. Wool and cotton are poor conductors of heat energy, so they stop our body heat from escaping.

Eat up!

Science explains why we need to eat food every day. Food contains stored energy that we need in order to move, grow, and keep warm.

On the move

Science explains how we can use a bicycle to travel quickly. The grip of the tyres produces friction between the wheel and the road, which helps to push the bike along.

Sun power

Science explains why a plant grows when we put it on a windowsill. Plants use energy from sunlight to produce food, which gives the plant energy to grow.

Time for bed

Science explains why the Moon stays in the sky. The force of gravity from the Earth pulls on the Moon, keeping it in orbit around our planet.

Ball games

Science explains why a ball moves when we kick, throw, or hit it. Our arms and legs produce a force that propels the ball forwards.

Science facts and figures

The world of science is full of amazing things. Here are some weird and wonderful facts and figures that you may not know.

40 litres (70 pints) is the **total amount of water** in the body of an adult man of average weight.

This is enough cement to make 5,000 of Egypt's Great Pyramid of Giza.

4 billion tonnes (4.4 billion tons) of **cement are made** in the world **every year**.

A **humpback whale's song** can **travel** more than **2,600 km** (1,600 miles) through water.

23

English scientist Isaac Newton was only 23 years old when he first described the force of gravity.

299,792

kilometres per second (186,282 miles per second) is the speed at which light travels. If you could move this fast, you could go around the Earth 7.5 times in one second.

The Eiffel Tower in Paris, France, "grows" by 15 cm (6 in) in summer as the iron it is made of expands.

97.2% **of all water** on Earth is **undrinkable** because it is salt water.

The number of times lightning hits the ground in the USA every year.

25 million

600 volts is the **shocking amount of electricity an electric eel** can produce. That is enough to kill any other fish nearby.

352,000 different species of flowering plant are alive in the world today, and there may be many more that are yet to be discovered.

10 mm (½ in) is the distance a maglev (magnetic levitation) train is lifted above the track using magnets.

Glossary

Here are the meanings of some words that are useful for you to know when learning about science.

attract When two things pull towards each other

bacteria Tiny creatures that live everywhere on Earth, such as inside food, soil, or even in the human body

boil When a liquid is heated to a temperature at which it bubbles and turns into a gas or vapour called steam

brittle Easily snapped

carnivore Animal that eats only meat

characteristics Certain qualities or features that things, animals, plants, or objects have

circuit Loop that an electric current travels around

condensation When a gas cools and becomes a liquid. Often seen as droplets of water that form on cold surfaces, such as windows

conductor Substance that allows heat or electricity to pass easily through it

echo Sound that has bounced off a surface and been sent back in the direction it started from

electricity Type of energy, which can be used to power household items

evaporation When a liquid is heated and changes into a gas or vapour

evolution Process where living things change, over many generations, to become new species

force Push or pull that causes things to start moving, move faster, change direction, slow down, or stop moving

friction Force created when two surfaces rub or slide against each other

gas State of matter with no fixed shape, such as air, that fills any space it is in

gravity Invisible force that pulls objects towards each other

groundwater Water found beneath the Earth's surface

herbivore Animal that eats only plants

invertebrate Animal that does not have a backbone

insulator Substance that does not allow heat or electricity to pass easily through it

The force of the tennis racket hitting the ball causes the ball to change direction.

irreversible Change that cannot be undone

laboratory Place where scientists carry out experiments

light Type of energy that allows humans and other animals to see

liquid State of matter that flows, and takes the shape of any container it is in

magnetism Invisible force that is created by magnets, which then pull certain metals towards them

mass Amount of matter that is in an object

material Substance that can be used to make things. It can be natural or made by humans

matter Stuff that all things are made of

melt When a solid is heated and becomes a liquid

mixture Combination of more than one type of thing

nutrients Food or substance that gives a living thing the energy or chemicals that it needs to live, grow, and move

omnivore Animal that eats both plants and meat

organ Body part that has a certain job, for example, the heart

organism Living thing

reflect When light or sound is bounced off a surface

repel When two objects push away from each other

reproduce To have young

reversible Change that can be undone

rust Reddish-brown crystals that form on iron and steel when they come into contact with water and oxygen

shadow Formed when light rays are blocked by a solid object

solid State of matter that holds its shape

solution Mixture that is created when a solid dissolves in a liquid and disappears

sound Form of energy that is produced when objects vibrate, or shake

Steel paperclips that have been attracted to a horseshoe magnet.

spectrum Range of something, for example, range of colours in a rainbow

temperature Measure of how hot or cold things are

vertebrate Animal that has a backbone

weight Amount of the force of gravity that acts on an object, making it feel heavy. The more mass something has, the larger the force of gravity on the object, and the heavier it feels

Index

Acknowledgements

The publisher would like to thank the following people for their assistance in the preparation of this book: Kathleen Teece, editorial assistant, Lucy Sims, designer, Emma Hobson, designer, Surya Sarangi, picture research, Alexandra Beeden, proofreader, Helen Peters, indexer, Dan Crisp, illustrator, and Lorraine Johnson for photography. The publishers would also like to thank Dr Suze Kundu of Imperial College London, UK, for the Meet the expert interview.

The publisher would like to thank the following for their kind permission to reproduce their photographs:

(Key: a-above; b-below/bottom; c-centre; f-far; l-left; r-right; t-top)

2 Alamy Images: Kim Karpeles (bl). **Corbis:** Hank Grebe / Purestock / SuperStock (bc). **3 Alamy Images:** Olga Khomyakova (bl). **4 Dorling Kindersley:** Natural History Museum, London (clb, fclb). **6 Corbis:** Momatiuk - Eastcott (tr); Dave Reede / All Canada Photos (b). **7 Corbis:** (tl). **8 Corbis:** 2 / Travelif / Ocean (cb). **8-9 Dreamstime. com:** Juhku. **10 Alamy Images:** Fotograferen.net (br). **Corbis:** 167 / Stacy Gold / Ocean (tl); India Picture (tr). **13 Corbis:** Chris Clor / Blend Images (crb/rusty). **Getty Images:** Fahim Shafayat Rahman / Moment Open (cra). **15 Corbis:** EPA (tr). **Dorling Kindersley:** Tim Draper / Rough Guides (tc). **16-17 Corbis:** Andrea Francolini. **17 Alamy Images:** Leroy Francis / hemis.fr (tr). **19 NASA:** JSC (tr). **20-21 Alamy Images:** David Wall (t). **22 naturepl.com:** David Pattyn (crb, bl). **23 Corbis:** Kirk Norbury / incamerastock (cla/squirrel); Roger Ressmeyer (tr). **NASA:** Joel Kowsky (cra). **naturepl.com:** David Pattyn (br, clb). **24 Corbis:** 145 / Steven Puetzer / Ocean (tr). **25 Alamy Images:** Kim Karpeles (clb). **Corbis:** Missen / RooM the Agency (tr). **27 Corbis:** Brett Stevens (bl). **29 Corbis:** robertharding (bl). **Dorling Kindersley:** NASA (tr). **30 Alamy Images:** matthiasengelien.com (tl). **31 NASA:** (cb). **34-35 Alamy Images:** Mike Grandmaison / All Canada Photos. **36 123RF.com:** Steve Allen / steveallenuk (br); Dimitar Marinov / oorka (bl); skylightpictures (tl); scalatore1959 (tr). **37 Dreamstime.com:** Monkey Business Images / Monkeybusinessimages (tl). **Getty Images:** Robbie George / National Geographic (r). **39**

Alamy Images: Leo Kanaka (fbr). **Dorling Kindersley:** The National Music Museum (crb). **Getty Images:** Nash Photos / Photographer's Choice RF (tr). **42 Ed Prosser:** (tr). **42-43 123RF.com:** Ekasit Wangprasert / bankerwin (b). **44-45 Dreamstime.com:** Geert Weggen / Geertweggen. **45 Dreamstime.com:** Yauheni Krasnaok / krasnaok (tr). **46 Alamy Images:** Zoonar GmbH / Peter Himmelhuber (tc). **Corbis:** Robert Pickett (tl). **46-47 Alamy Images:** Olga Khomyakova. **47 Corbis:** AgStock Images (r); Houin / BSIP (tr). **48 Corbis:** Dave Michaels (cl); Thijs van den Burg / NIS / Minden Pictures (cr). **49 Alamy Images:** Don Johnston_PL (tr); Nurlan Kalchinov (cr). **Corbis:** Science Photo Library: Dr. Keith Wheeler (br). **50 Corbis:** 145 / Ocean (c); Nature Picture Library (br). **Dorling Kindersley:** Natural History Museum, London (tl, tc). **Fotolia:** StarJumper (cl); Tujian (ca). **50-51 Fotolia:** Tujian. **51 Corbis:** Nature Picture Library (bc, br). **52-53 Corbis:** Hank Grebe / Purestock / SuperStock. **54 Alamy Images:** Pictorial Press Ltd (cr). **Corbis:** Bettmann (clb, ftl). **Dorling Kindersley:** Natural History Museum, London (tl). **55 123RF.com:** Narongsak Yaisumlee / studio306 (ftr). **Alamy Images:** Everett Collection Historical (cb). **Corbis:** (tl); Oscar White (clb); Underwood & Underwood (c). **Fotolia:** Dario Sabljak (tr). **56 Corbis:** Monkey Business Images (cr); Ian Nolan / cultura (bl). **57 Corbis:** 2 / Mark Lund / Ocean (tr); Leander Baerenz / Westend61 (cl). **Dorling Kindersley:** Getty Images / Photographer's Choice RF (bl). **Dreamstime.com:** Gregsi (br). **58 Getty Images:** M Sweet / Moment (cr); ScPhotographie / Photographer's Choice (b). **iStockphoto.com:** mstay (cl). **59 Alamy Images:** Bernd Mellmann (br). **Corbis:** 145 / Jurgen Reisch / Ocean (ca). **60 Alamy Images:** Mike Grandmaison / All

Canada Photos (tl). **Corbis:** Leander Baerenz / Westend61 (br). **62 Alamy Images:** David Wall (tl).

Cover images: *Front:* Corbis: Dave Michaels br; **Dreamstime.com:** Jakub Gojda / Jagcz clb; *Back:* **123RF.com:** Dimitar Marinov / oorka tl; **Alamy Images:** Everett Collection Historical cra; **Corbis:** 145 / Steven Puetzer / Ocean cr, AgStock Images tr; *Spine:* **Getty Images:** Nash Photos / Photographer's Choice RF; *Front Flap:* **123RF.com:** Ekasit Wangprasert / bankerwin bl/ (back); **Alamy Images:** Kim Karpeles c/ (front); **Corbis:** 145 / Ocean cla/ (back), Hank Grebe / Purestock / SuperStock br/ (back); **Dorling Kindersley:** The National Music Museum cr/ (back); **Fotolia:** Tujian cra/ (back); **Getty Images:** Nash Photos / Photographer's Choice RF tr/ (back); **naturepl.com:** David Pattyn tc/ (front); *Back Flap:* **Dorling Kindersley:** Gary Ombler / The University of Aberdeen cl/ (front), Tim Parmenter / Natural History Museum, London crb/ (front); **NASA:** JSC clb/ (front).

Endpapers: *Front:* **Alamy Images:** Georgios Kollidas (gravity); Kumar Sriskandan (columns). **Dorling Kindersley:** Museo de Sitio Huaca Pucllana (bricks); The Science Museum, London (wheel). *Back:* **Alamy Images:** Everett Collection Historical (penicillin); GL Archive (lovelace); Pictorial Press Ltd (einstein). **Corbis:** Bettmann (darwin, heart transplant); Najlah Feanny / Corbis Saba (cloning). **Dorling Kindersley:** Gary Ombler / Whipple Museum of History of Science, Cambridge (battery); The Science Museum, London (lightbulb). **Getty Images:** CBS Photo Archive / CBS (goodall).

All other images © Dorling Kindersley
For further information see:
www.dkimages.com

My Findout facts:

Great moments in science

The first battery was a stack of zinc and silver discs, called a "voltaic pile".

Charles Darwin
Englishman Charles Darwin published *On The Origin of Species,* which describes his theory of evolution.

Radioactivity
Polish scientist Marie Curie and her French husband, Pierre Curie, discovered two radioactive substances called polonium and radium.

Flasks and test tubes used by Alexander Fleming to develop penicillin.

Batteries
Italian physicist Alessandro Volta made the first battery.

1799	1842	1859	1879	1898	1915

← Continued from front of book

Electric light bulb
American inventor Thomas Edison created an electric light bulb that was cheap to make and lasted a long time..This meant that homes and cities could have lights that were safe to leave on all night.

Albert Einstein
German physicist Albert Einstein published a paper called *The Foundation of the General Theory of Relativity,* which explained his theories about energy, mass gravity, space, and time.

Ada Lovelace
English mathematician Ada Lovelace wrote the first program for a very early version of a computer.

Thomas Edison formed the Edison Electric Light Company, which made this early version of an electric light bulb.